Microsoft®

What's New in Office™ 2007

NEVADA LEARNING SERIES™

QUICK REF

Getting to Know Office 2007

Microsoft Office 2007 has been updated to make frequently used commands and functions easier to find. The Ribbon runs across the top of the window, replacing the menus and toolbars found in previous versions of the Microsoft Office suite. Commands are grouped in the Ribbon under tabs that are organized by task, making it easier to locate the feature that you need. The Ribbon is designed to minimize clutter by only displaying commands relevant to the work being done.

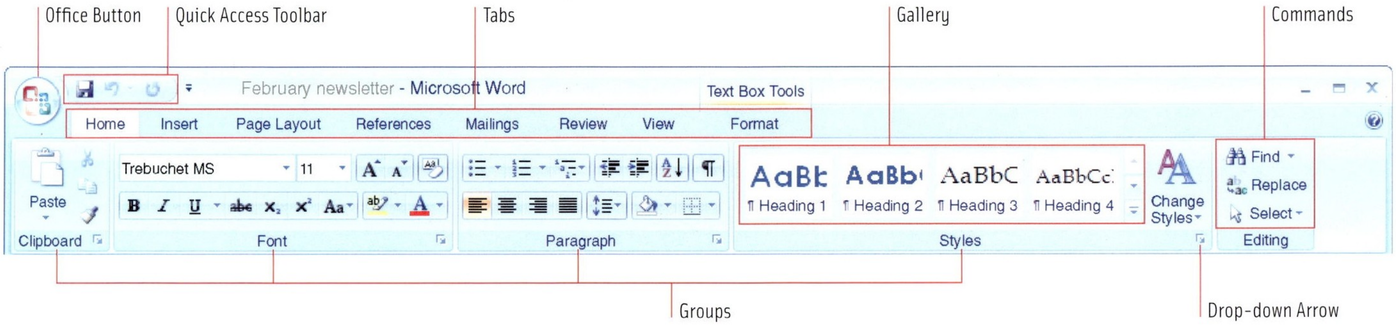

Displaying the Ribbon

The Ribbon is displayed by default when you open Microsoft Word 2007, Excel 2007, or PowerPoint 2007. You can minimize the Ribbon to view more of your screen.

To minimize the Ribbon: double-click anywhere on the tabs. Alternatively, right-click anywhere on the Ribbon and choose **Minimize the Ribbon**.

To restore the Ribbon: click any tab to activate it temporarily or double-click to restore the Ribbon.

Ribbon Components

There are three basic components to the Ribbon.

Tabs: Tabs are arranged across the top of the Ribbon. Each tab represents core tasks in a given program. Contextual tabs are tabs that only appear in certain contexts.

Groups: Groups are sets of related commands, displayed on tabs.

Commands: Commands are buttons, menus, or boxes where you enter information. They are grouped according to the type of tasks that are available at a given time.

Office Button

The Office Button replaces the File menu in previous versions of Microsoft Office. It contains commonly used commands such as Open, Save, and Print.

To create a new document, presentation or workbook: click New and double-click **Blank Document/Presentation/Workbook**. Alternatively, press CTRL+N.

Tip: For more information on the Office Button, see page 1 of the Word 2007, PowerPoint 2007, or Excel 2007 Quick Reference Guides.

Quick Access Toolbar

You can customize the Quick Access Toolbar to display the commands that you use most frequently.

To customize the Quick Access Toolbar: right-click a command icon on the Ribbon (*e.g.* **B** from the Home tab), and choose **Add to Quick Access Toolbar**.

To move the Quick Access Toolbar: right-click anywhere on the Ribbon and choose **Show Below the Ribbon**. To move the Toolbar back above the Ribbon, choose **Show Above the Ribbon**.

KeyTips

The commands on the Ribbon, Office Button, and Quick Access Toolbar can be executed by using KeyTip access keys. The access keys are only unique within the selected tab.

To use KeyTips: press ALT on your keyboard. Access keys are activated over the features that are currently available. For example, pressing ALT+H displays access keys in the Home tab, ALT+N displays access keys in the Insert tab, *etc.* Press the letter over the feature that you want to use.

Previewing

Galleries in the Ribbon contain predefined styles, effects, and table formats. Use live previews of the formatting in your document before you make a selection.

To preview a choice: rest the mouse pointer on an option to see a preview. Click on it to apply it to your document.

Table Of Contents

Formatting Documents

You can easily apply formatting changes to your documents with Word's new Quick Style sets, building blocks, and style galleries. When your document is complete, you can convert it to PDF file format.

Working with Styles

You can quickly format headings and paragraphs using Quick Style sets that contain styles designed to work together. Word generates a table of contents based on the heading styles you use.

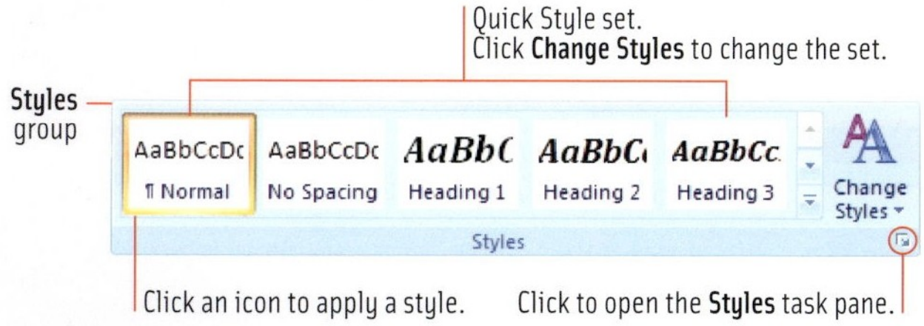

To select a Quick Style set: click the Home tab, click **Change Styles** in the **Styles** group, and then select a style set from the list (*e.g.* **Distinctive**).

To apply a style: select text and click a style icon. To see more style options, click to the right of the style icons to expand the Quick Styles gallery.

To create a new style: click at the bottom right corner of the **Styles** group. Click in the lower left corner of the **Styles** task pane and create a style in the **Create New Style from Formatting** dialog box. Click **OK**.

To change an existing style: right-click a style icon and select **Modify**. Make your changes in the **Modify Style** dialog box. Click **OK**.

Using Building Blocks and Style Galleries

Word's new document building blocks and expanded galleries of style formats, table formats, and list formats, allow you to quickly redesign your documents.

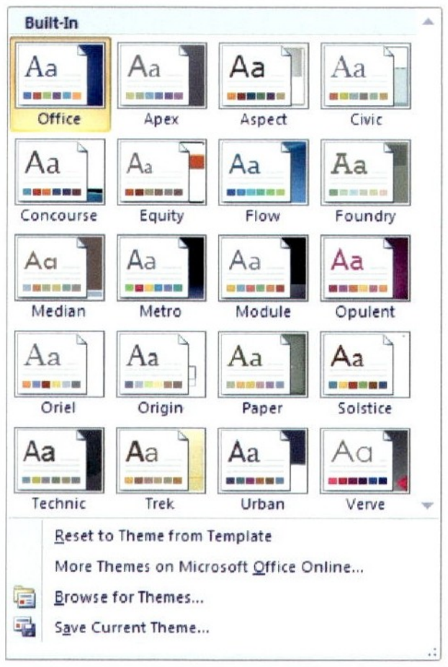

Before you make a style change in your document, the galleries of formatting choices provide you with a visual preview of what the style will look like.

Word's new building block feature allows you to add preformatted content to your documents. You can select from a wide range of preformatted cover pages, pull quotes, and add headers and footers to enhance the appearance of your document. Shown at right are some of the document themes.

To insert building blocks in your document: position the cursor in your document. Click the Insert tab, and then click **Quick Parts** ▶ **Building Blocks Organizer** in the **Text** group. In the **Building Blocks Organizer** dialog box, select one of the pre-formatted items (*e.g.* a cover page) and click Insert.

Converting Documents to PDF File Format

Word 2007 supplies an easy-to-install free add-in that enables you to convert Word documents to PDF file format. This capability is useful if you want to reduce large documents to a more manageable file size so that they can be easily distributed via e-mail. Sending a document as a PDF file also ensures that your recipients will be able to open it, even if they are not using Word.

To enable you to save your document as a PDF file: install the Microsoft add-in for PDF conversion. For instructions, click ▶ **Save As** ▶ **Find add-ins for other file formats**. Once installed, use the **Save As** dialog box to select the PDF file type.

Reviewing and Protecting Documents

Word's Compare command and Document Protection on the Review tab makes collaborative review and editing more manageable. During the editing process, you can compare different versions of your document. Before you distribute the document, you can protect it by removing personal information and restricting changes.

Comparing Two Documents

The **Compare** command on the Review tab arranges documents in the same window for easy comparison (see illustration below). This feature is useful during editing when you might want to work with several versions of the same document. Once this view is active, you can scroll both documents simultaneously.

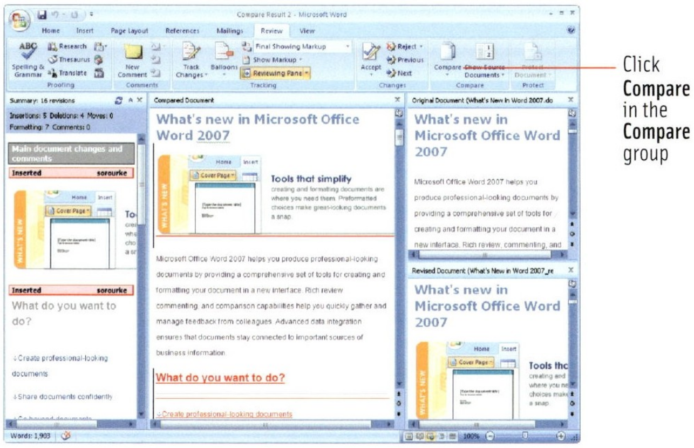

To compare changes in two versions of a document: click **Compare** ▶ **Compare** in the **Compare** group. In the **Compare Documents** dialog box, locate the appropriate documents using the **Original Document** and **Revised Document** areas. If either document version has tracked changes, these will be displayed in red text or markup balloons.

To merge changes from multiple reviewers: click **Compare** ▶ **Combine** in the **Compare** group. In the **Combine Documents** dialog box, locate the appropriate documents using the **Original Document** and **Revised Document** areas. Click **OK**. Repeat the procedure to merge the combined document with each subsequent revised document.

Enabling Contextual Spelling

You can now spend less time proofreading and editing your documents; the spell checker in Word 2007 can correct words in context. For instance, it can recognize and replace *there* with *their*.

To enable contextual spelling: click ▶ Word Options and select **Proofing** from the list in the **Word Options** dialog box. Make sure that the **Use contextual spelling** box is checked.

Protecting Your Documents

Before circulating a draft document for review, you can set restrictions that limit text changes or enforce the use of specific text styles. These restrictions prevent reviewers from making unwanted edits or style changes. You can also check your document for hidden metadata, personal information, or stored content.

To restrict changes to a final version of the document: click ▶ **Prepare** ▶ **Mark as Final**, and then click **OK** to save the document.

To find hidden personal information and metadata in a document: click ▶ **Prepare** ▶ **Inspect Document**. If prompted to save your document, click **Yes**. In the **Document Inspector** dialog box, check or clear the boxes beside the document inspection options, and then click Inspect. If necessary, click Remove All beside found items.

Viewing and Protecting Your Worksheets

You can make the content of your worksheets easier to view by adjusting the magnification settings or by switching to Page Layout view. You can also protect the information in your worksheets by setting editing restrictions and removing personal information.

Adjusting Worksheet View Settings

The new zoom slider appears on the status bar at the bottom of the worksheet. Instead of having to choose a new magnification percentage from a drop-down menu, you can quickly zoom in and out on the worksheet using the slider.

To zoom: at the bottom-right of the worksheet, click the zoom slider 100% and drag it right to zoom in or left to zoom out. Alternatively, click on the zoom value, choose a magnification option (*e.g.* **50%**) in the **Zoom** dialog box, and click **OK**.

Making Changes to Worksheets in Page Layout View

Excel 2007 includes a new view called Page Layout, which allows you to see how your worksheet will print and to make final layout changes to your document.

To switch to Page Layout view: click the View tab, and then click **Page Layout** in the **Workbook Views** group. Alternatively, click the Page Layout View icon on the status bar at the bottom-right of the worksheet.

To hide or display the rulers: in **Page Layout View**, on the **View** tab, in the **Show/Hide** group, check or clear the **Ruler** box.

To add a header or footer: in **Page Layout View**, on the **Insert** tab, in the **Text** group, click **Header & Footer**. The pointer is automatically placed inside the header text box; to switch to the footer text box, click **Go to Footer**. Type the text you want to insert.

To set page margins: in **Page Layout View**, on the Page Layout tab, in the **Page Setup** group, click **Margins**, and then choose **Normal**, **Narrow**, or **Wide**. For more options, click **Custom Margins**, and then, on the Margins tab, choose the margin sizes you want and click **OK**.

Protecting Your Worksheets

Before you share a worksheet with others, you can set restrictions to prevent them from making unwanted changes. You can also check the worksheet for hidden metadata and personal information.

To protect cells against changes: click the Review tab, and then click **Protect Sheet**. Check the **Protect worksheet and contents of locked cells** box, and then check the boxes beside the actions you want users to be able to perform. Type a password and click **OK**. Retype the password and click **OK**.

To protect worksheets against deletion: click the Review tab and click **Protect Workbook** ▶ **Protect Structure and Windows** (see below). Check the **Structure** box, and then type a password and click **OK**. Retype the password and click **OK**.

Click **Protect Workbook** in the Changes group

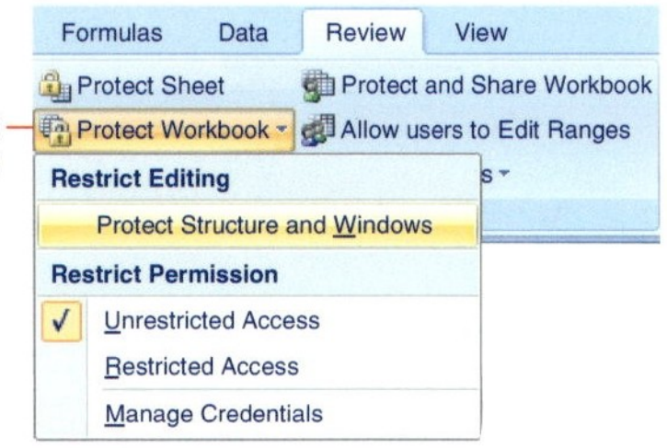

To scan the document for personal information that may be hidden from view: save your work and choose ▶ **Prepare** ▶ **Inspect Document**. Choose options in the **Document Inspector** dialog box and click Inspect.

To encrypt the document and prevent unauthorized use: choose ▶ **Prepare** ▶ **Encrypt Document**. Type a password and click **OK**. Retype the password and click **OK**.

Adding Formatting Effects and Charts

You can add consistent colors, fonts, and fill effects to your documents using Themes. You can also highlight information by adding colors or data bars to cells and by inserting charts.

Applying Themes

You can quickly give your document a consistent, professional look by applying standard sets of colors, fonts, and fill effects called Themes. Themes can be customized to match corporate styles, saved for future use, and shared among colleagues. Themes can also be shared across the Office 2007 suite.

To apply a theme: click the Page Layout tab, and then click **Themes**. Move your mouse over the different themes in the list and your worksheet will preview the current theme. Click on a theme to select it.

To modify an existing theme: click the Page Layout tab, and then click either Colors, Fonts, or Effects in the **Themes** group.

To save a theme: click the Page Layout tab, and then click **Themes** ▶ **Save Current Theme**.

Identifying Data with Conditional Formatting

Conditional Formatting automatically adds colors or bars to cells depending on the values the cells contain. This makes it easier for you to identify data that meet certain requirements (such as numbers that fall into a specified range or exceed a critical value) and to see patterns and trends on the worksheet.

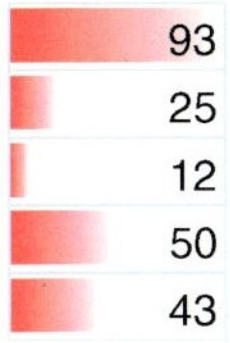

To visually identify data using bars or colors:

1. Select the cells you want to format, click the Home tab, and then click **Conditional Formatting** in the **Styles** group.
2. In the **Conditional Formatting** drop down list, choose **Data Bars** or **Color Scales**. Data Bars or Color Scales now appear in each cell, making it easy to find high, low, and average values.

Creating and Customizing Charts

Charts in Excel 2007 are easier to create and customize. Select the data and choose a chart style from the Ribbon to quickly produce your chart. The chart type palettes, which display when you click a chart style, previews what your chart will look like.

To create a chart: select the range of data on which to base your chart. Click the Insert tab and select the appropriate chart for your data from the **Charts** group.

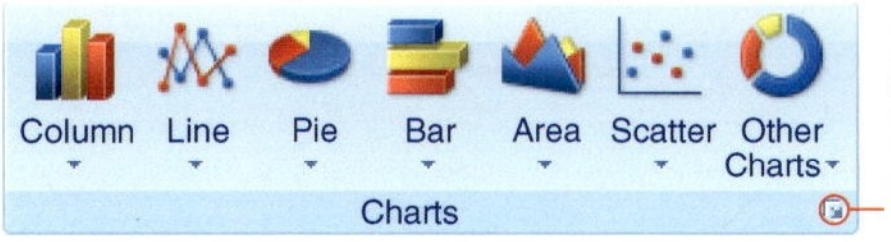

Choose a chart type from the Ribbon, or click here for additional options.

To add a title to a chart: select the chart, click the Layout tab, click **Chart Title**, and select a title position. Click the chart title and enter a title.

To add a title to an axis: select the chart, click the Layout tab, click **Axis Titles**, and select an option. Click the axis title and enter a title.

To change the layout of a chart: select the chart, click the Design tab, and select an appropriate layout from the **Chart Layouts** group. To view more chart options choose.

To change the format elements of a chart: select the chart element you wish to change, choose the Format tab, and make your changes.

To save a chart as a template: select the chart, click the Design tab, and click **Save As Template** from the **Type** group.

To use a chart in another Office application: select the chart, click the Home tab, and click **Copy**. Paste your chart in your other Office application.

Reviewing Your Presentation

PowerPoint 2007 has features to help you review your presentation before showing it or sending it to others. You can remove information not meant to be viewed by others and disable changes to the final version.

Using the Document Inspector

Before sending your files out to your contacts, you can have them inspected for personal information, comments, *etc.* that you do not want to include. The Document Inspector checks your file, lets you know if any such information is included, and then gives you the opportunity to have it removed. You can also mark a file before sending it out so that others know it is the final version.

1. Click ▶ **Prepare** ▶ **Inspect Document**. If you are prompted to save the file, click **Yes**.
2. In the **Document Inspector** dialog box, select the content you want to check (*e.g.* **Comments and Annotations** or **Invisible On-Slide Content**) and click Inspect. The inspection results will flag any concerns with an exclamation mark !.
3. Click Remove All to remove the information flagged by the inspector, and then click Close.

Marking a Presentation as Final

Marking presentations as final makes them read-only; editing and proofing features are disabled.

1. Choose ▶ **Prepare** ▶ **Mark as Final**. Click **OK** to confirm the status change, save the presentation, and then click **OK** again to close the message.
2. The Marked as Final icon will display in the status bar.

Note: To remove the Marked as Final status, choose ▶ **Prepare** ▶ **Mark as Final**.

Running Your Presentation

You can run your presentation using two monitors: one to show your audience and one in Presenter view to display thumbnails, speaker's notes, navigation controls, and preview text. By choosing the resolution in which to show your presentation, you can optimize speed or visual detail.

Using Presenter View

1. Click the Slide Show tab and check the **Use Presenter View** box ☑ in the **Monitors** group.
2. In the **Show Presentation On** drop-down list, click the icon for the monitor you want to display to your audience.
3. Your second monitor will display the Presenter view.

To choose the resolution for your presentation: in the **Monitors** group of the Slide Show tab, use the **Resolution** drop-down list to choose how you want to display your presentation (high resolution will have greater visual detail but performance will be slower).

To show a presentation: press F5 or click the Slide Show tab and select **From Beginning** in the **Start Slide Show** group. When the slide show starts, click the mouse or press the spacebar to advance the slides. Use the presentation toolbar in the bottom left corner or right-click to show menu options.

To add annotations during a presentation: click and choose a pen type from the menu (*e.g.* **Highlighter**), then click and drag to mark up the slide. To turn off the annotation tool, choose **Arrow** from the same menu.

To erase an annotation: click , choose **Eraser**, and then click on the annotation or select **Erase All Ink on Slide**.

To switch to a black screen: right-click and select **Screen** ▶ **Black Screen**. C

once or press the spacebar to restart the presentation.

To end the presentation: press ESC, or click and choose **End Show**.

Designing Your Presentation

SmartArt allows you to take existing text and turn it into a dynamic diagram with professional looking graphics that are laid out and formatted for you.

Using SmartArt

SmartArt templates can be used for a variety of needs, including organizational charts and diagrams, and reflect slide themes.

To insert a new SmartArt Graphic:

1. Click the Insert tab and click **SmartArt** from the **Illustrations** group.
2. In the **Choose a SmartArt Graphic** dialog box (see below), choose a category from the left side and click on any SmartArt thumbnail to see a description on the right (*e.g.* click **Hierarchy**, and then click a Chart to see a description).

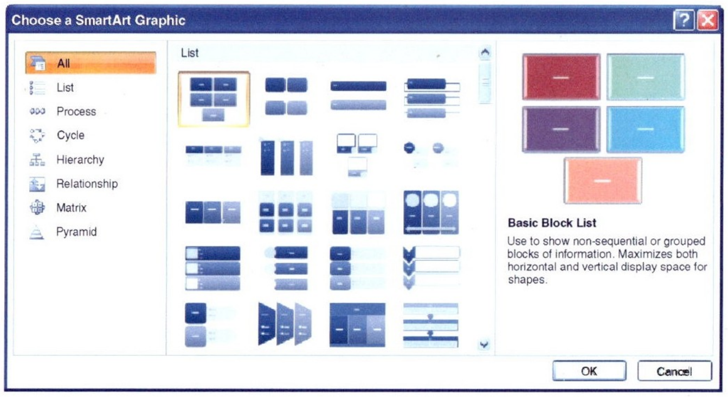

3. Click **OK** to insert the selected SmartArt graphic into your slide. The **SmartArt** Design tab displays.
4. Click on the SmartArt to enter text in the placeholders, or click the arrows on the left side to display the Text Pane and type your text there. Click ✕ to close the Text Pane.
5. Use the **SmartArt Tools** and Design and Format tabs to format your graphic.

To create a SmartArt graphic from existing text: select a text box, click the Home tab, and then click **Convert to SmartArt Graphic** in the **Paragraph** group. Use the gallery to preview the effects and then click one to apply it.

Note: When you convert the text to a SmartArt graphic, you may lose some of its formatting (*e.g.* font size, font color).

To add shapes to the SmartArt graphic: select a shape in the graphic, click the **SmartArt Tools** Design tab, click **Add Shape** ,and then choose an option in the **Create Graphic** group.

To make a SmartArt graphic shape invisible: select the shape, click the **SmartArt Tools** Format tab, and then, under **Shape Styles**, click **Shape Fill** and choose **No Fill**. From the **SmartArt Tools** Format tab, under **Shape Styles**, click **Shape Outline** and choose **No Outline**.

Note: As you add shapes to the graphic all other elements get smaller. SmartArt can only be animated as a group in the order (or reverse order) in which the elements appear.

To change the colors of a SmartArt graphic: select the SmartArt graphic, click the **SmartArt Tools** Design tab, and then click **Change Colors** in the **SmartArt Styles** group. Use the gallery to preview the choices and click one to apply it.

To remove changes made to the SmartArt graphic: click the **SmartArt Tools** Design tab, and then click **Reset Graphic**.

Note: To remove the changes for only one shape, right-click the shape and

9781553748564

ISBN: 978-1-55374-856-4

:s.com

io

Canada